Sukhoi T-4 Sotka

The Soviet Mach 3+ Hypersonic Missile Carrier/Airborne Reconnaissance System

HUGH HARKINS

Sukhoi T-4 Sotka

The Soviet Mach 3+ Hypersonic Missile Carrier/Airborne Reconnaissance System

© Hugh Harkins 2018

Centurion Publishing

United Kingdom

ISBN 10: 1-903630-75-4
ISBN 13: 978-1-903630-75-4

This volume first published in 2018

The publisher and author would like to thank all organisations and services for their assistance and contributions in the preparation of this volume: JSC Myasishchev Design Bureau (EMZ); JSC Sukhoi; Ministry of Defence of the Russian Federation; NASA Glen Research Centre; NPO Saturn; OJSC Dubna Machine Building Plant; OJSC Tupolev; Russian Federation Air Force Museum – Monino; TsAGI – Central Aerodynamic Institute; United Engines Corporation; and the US DoD

CONTENTS

INTRODUCTION

The intent of this volume is to detail the Soviet Union's Mach 3+ capable Sukhoi T-4 Sotka hypersonic missile carrier/Airborne Reconnaissance System designed and developed in the late 1960's and early 1970's as a conventional and nuclear counter to enemy aircraft carrier battle groups threatening Soviet territory. Prior to the programs cancellation in 1975, a single T-4 aircraft entered flight testing.

The volume covers the genesis, design, development of the T-4 and the X-45 hypersonic missile armament and an overview of the flight tests of the former, which were conducted in the period August 1972 through mid-January 1974 when the program was suspended. Discussions of the various unofficial infighting within Sukhoi and disagreements between the heads of the various design bureaus of Sukhoi, Tupolev and Myasishchev are avoided – these are mere tangents that are not conducive to the aim of the volume and divert from the detail of the T-4 Sotka aircraft program as designed, built and flown, and the X-45 armament intended for the Sotka.

SUKHOI T-4 SOTKA – THE SOVIET MACH 3+ HYPERSONIC MISSILE CARRIER/AIRBORNE RECONNAISSANCE SYSTEM

The design that emerged as the Sukhoi T-4 hypersonic missile carrier/airborne reconnaissance system was born out of Soviet studies for a survivable counter to NATO (North Atlantic Treaty Organisation) aircraft carrier battle groups. Such groups could threaten the Soviet Union with air attack from strike aircraft armed with either conventional or nuclear weapons. From the earliest days of the Cold War, Soviet planners were very conscious of the overwhelming naval superiority that NATO possessed. Although plans were initiated to provide a future Soviet navy that could counter NATO in selected areas, the Cuban Missile Crisis (Soviet termed Caribbean Crisis) of October 1962 had demonstrated that the Soviets were unable to decisively challenge NATO aircraft carrier, either strike or ASW (Anti-Submarine Warfare), battle groups at oceanic distances from Soviet territory. At the time of the 1962 crisis, measures to redress NATO's surface fleet superiority were in progress. This included enhancement of submarine launched ASCM (Anti-Ship Cruise Missile) capability and air launched ASCM capability of Soviet Naval Aviation. In the latter regard, near term measures included enhancements to existing naval strike aircraft such as the Tupolev Tu-16K subsonic missile carrier/bomber, which could be armed with various ASCM. Longer term, it was proposed to introduce a highly survivable maritime strike aircraft with very high speed carrying high supersonic air to surface missiles that would be immune to surface defences and be capable of neutralising the threat posed to the Soviet homeland by aircraft carrier-borne strike aircraft.

A not insignificant secondary benefit from neutralising NATO carrier battle groups was the removal of NATO's unquestioned air supremacy on the world's oceans, particularly the North Atlantic and Pacific. Removal of NATO carrier borne fighter aircraft, through destruction or disabling their operating platforms, would allow Soviet maritime patrol/ASW aircraft to, with a higher degree of safety than would hitherto be the case, patrol oceanic areas suspected to be operating areas for NATO SSBN (ballistic missile armed nuclear powered submarines) or, indeed, to more safely hunt NATO SSN (nuclear powered attack submarines) in oceanic areas being used for the operation of Soviet ballistic missile submarines.

The Sukhoi T-4 prototype, code 101, at the LII airfield at Zhukovsky in summer 1973.
Sukhoi

Since the early 1950's, there had been a steady progression of Soviet aviation-borne anti-ship capability. The prototype of the Tu-16, which was the first Soviet operational jet powered long range bomber, had flown on 27 April 1952 (pilot N. Rybko). Production was authorised in December that year, commenced in 1953 and ended in 1962. As well as bomber and missile carrier roles the Tu-16 was developed for support roles including reconnaissance, electronic warfare and airborne tanker. The last of the Tu-16's, a support role variant, was taken off combat duty with the Russian Federation in 1993 (MODRF). The premier anti-ship variant, the 75800 kg take-off weight Tu-16K-11-16 missile carrier/bomber, had a maximum speed of 1050 km/h, an operating ceiling of 15000 m, an operational range of 7800 km and could be armed with two KSR-2, KSR-5 (Tu-16K-26/16-10-26) or KSR-11 ASCM. The deployment of the Tu-16K in 1959 was a significant leap forward in capability, but it was a step short of the desire to introduce a highly survivable missile carrier platform that would not only be less vulnerable to ship based and fighter aircraft defences, but would also be able to respond, in a more timely fashion, to time critical targets. This requirement led to deployment of the 92000 kg take-off weight Tu-22/K, which was the first Soviet operational long range bomber capable of supersonic speeds – maximum design speed was 1640 km/h, operational ceiling was 13500 m, operational range was 5650 km and the aircraft could be armed with a single X-22M/MA series ASCM. The 122000 kg take-off weight Tu-22M emerged, in the early 1970's with a maximum speed of 2000 km/h, an operating ceiling of 14000 m, an operational range of 7000 km and could be armed with three X-22 series ASCM. The X-22 series, N/HA, ASCM also armed the turboprop powered subsonic Tu-95K-22 long-range anti-ship missile carrier. While performance in the areas of speed and ceiling were considerably inferior to that of the Sukhoi T-4 design, the Tu-22M had similar range values to the Sukhoi aircraft. The X-22M/MA

4

ASCM was vastly inferior in overall performance and operational capability compared to the X-45 (Kh-45) hypersonic missile planned for the T-4 (the X-45 could, of course have been applied to the Tu-22M2/3 designs). The evolution from the Tu-16K to the Tu-22K saw crew levels drop from six to three. The Tu-22M required a crew of four whilst the T-4 required a crew of only two, this latter aircraft lacking the defensive cannon armament of the Tu-16K/Tu-22K/Tu-22M series as it was designed to fly too high and too fast for defensive fighter/interceptor aircraft to effect an interception.

Tupolev Tu-16K (top) and Tu-22K (above). US DoD/Tupolev

The Myasishchev M-50 supersonic strategic bomber prototype flanked by two Soviet jet fighter aircraft. USN HC

Design work on what would emerge as the T-4 commenced in 1961 at Sukhoi Design Bureau with a team led by chief designer N.S. Chernyakov. The aircraft was designed under the code 'article 100', projected weight being ~100 tons (Sukhoi). The year 1961 also saw the termination of flight tests of the Myasishchev M-50 supersonic strategic bomber prototype – the last flight of this aircraft being conducted on 9 July that year (the M-50 prototype had conducted its maiden flight with the crew of N.I. Goryainov and A.S. Lipko on 28 October 1959). The M-50 had been designed as a strategic bomber/missile carrier. In the latter role it would have carried a large strategic strike air to surface missile designated B-3 (V-3). Take-off weight of the M-50 was 210000 kg, maximum speed was 1950 km/h, ceiling was 16500 m and range was 7400 km. However, despite its considerable increase in speed and ceiling performance over the subsonic Myasishchev M-4 (the prototype had conducted its maiden flight on 20 January 1953) and the later 3M (the prototype had conducted its maiden flight on 27 March 1958) subsonic strategic (intercontinental) bomber (the 3M had a design maximum speed of 950 km/h, a ceiling of 11000 m and a design maximum range of 13000 km), performance was considered inadequate to ensure survival against the then perceived future air defence threat. The speed and ceiling performance were woefully short of that required to fill the maritime strike role for which the T-4 would be designed. Performance shortfalls aside, the M-50 would have been too heavy to be considered for the Soviet future maritime missile carrier role envisioned for the period of the late 1970's through the 1980s.

There is a high degree of uncertainty as to what specific reconnaissance roles would have been attributed to the planned maritime hypersonic missile carrier/reconnaissance aircraft. The Mach 3 speed would have made the design unattractive to the ASW/maritime patrol role undertaken by first generation Tu-95 and Tu-142 four engine turboprop aircraft (top). The Soviet Union had already conducted studies into multi-Mach capable reconnaissance aircraft such as the OKB-256 NM-1 (above), designer P.V. Tsybin, which was authorised by a government decree of 9 April 1956. This was initially to be powered by two D-21 direct-flow engines and be capable of Mach 3.2 speed with a 36 km ceiling, but work on the project ceased in 1962. US DoD/DMB

The operational requirement that led to the T-4 called for a platform capable of conducting 'reconnaissance, search and destroy missions on small-size, mobile and fixed offshore and onshore targets', for which an aircraft with an operational flight range of ~7,000 km was desired (Sukhoi). From the late 1950's through the 1960's Tupolev's UAV (Unmanned (Uninhabited) Air Vehicle) department was developing the series 121 and 123 missiles and a series of supersonic missile carrier/bombers beyond the Tu-22/K. Such studies would, in the second half of the 1960's, evolve into the design that ultimately emerged as the Tu-22M series. At the same time Tupolev was involved in design/development work on the Soviet SST (Supersonic Transport) that led to the Tu-144, it often being taken, that as Tupolev was involved in SST work, then Sukhoi could offer a variant of the T-4 for such an enterprise. However, in actuality, the T-4 was too small in overall dimensions and mass for such considerations on a serious level.

In the late 1950's, Tupolev was working on the 125 and 135 bomber projects, whilst Myasishchev was working on the M-52 bomber project. These designs were intended for fight regimes in the 2000-2500 km/h range, with aluminium alloys specified as the major construction materials. Myasishchev was also studying the M-56 to be constructed from titanium and steel alloys that would allow it to fly in the 3000 km/h speed range. This design, however, was projected as a heavy strategic missile carrier/bomber aircraft with a combat weight well in excess of 200 tons and an operational payload in the 9000-10000 kg range. Consequently, Myasishchev was not involved in the competition to develop the smaller ~100 ton missile carrier.

Myasishchev M-56 strategic missile carrier/bomber concept of 1960. Myasishchev

Three of the major design bureau participated in the competition – Sukhoi, Tupolev and Yakovlev. Tupolev may have appeared the favoured, even the logical, choice of design bureau to undertake the new bomber program since it was already advanced in the 135 bomber project. This would, however, have required a redesign to allow for the increased speed requirement of 3000+ km/h. Sukhoi was largely considered a fighter design house, although this was, in itself, a fallacy as it had conducted work on light bombers and ground attack aircraft in the past, although nothing approaching the size scale of the missile carrier project that was the subject of the competition. However, the bureaus participation in the program that led to the T-4 is not as unusual as is often proclaimed as Sukhoi was attempting to make inroads into the strike/attack realm, eventually developing the T-6 project that led to Su-24/M series of variable-geometry strike aircraft. Going back to the 1940's, Sukhoi had designed the Su-8 twin-engine armoured attack aircraft, the Su-12 training bomber and the Su-10 jet powered bomber. Much of the design requirements for the T-4 were more that of a large long-range strike aircraft than that of a strategic bomber, bringing it into the realm of projects within Sukhoi's planned portfolio.

In July 1961, a report from the Soviet Scientific and Technical Council provided an assessment of the competing designs from the respective design bureau of Sukhoi, Tupolev and Yakovlev. Sukhoi was the only design house to submit a concept that had the acceptable performance characteristics, including a maximum speed on 3000+ km/h, and to fall within the desired weight class – ~100 tons. Tupolev's project 135 was deemed unsuitable as its maximum speed, which was still projected at 2500 km/h, fell woefully short of the requirement for a 3000+ km/h speed at high altitude. Perhaps more damning to Tupolev's proposal was the fact that the projected weight of the design was around 190 tons, which was deemed well in excess, almost by half, of what was desired. The major benefits that the Tupolev design had brought to the table was a reduction in development costs at the expense of operational capability. Although the higher operational weight would allow the Tupolev concept to centre on a common airframe for the duel task of long-range maritime strike and strategic bombing against the continental United States, the gamble had failed as the Tupolev design was rejected.

The Yakovlev Yak-33 had a passing resemblance to the Myasishchev M-54, which itself bore external similarities to that of the Convair B-58 four jet Mach 2 capable bomber. The Yak-33, however, was designed for a maximum speed in the region of 3300 km/h (in excess of ~Mach 3.1 at high altitude). Despite meeting the speed requirement, the Yaovlev design, at 84 tons take-off weight, was considered below the specified 100 tons, resulting in a smaller aircraft that was deemed necessary for the planned hypersonic cruise missile carrier mission.

Both the Sukhoi and Yakovlev designs progressed to the next stage of the selection process, but, at the Scientific and Technical Council meeting held in September 1961, Tupolev, attempting to find a way back into the completion, put forward the Project 125, which it had specified as a direct Tu-22/K replacement. However, this design, with a maximum speed of 2500 km/h (the same as that of the 135), failed to meet the desired 3000+ km/h requirement and ultimately, like the 135 before it, fell by the wayside.

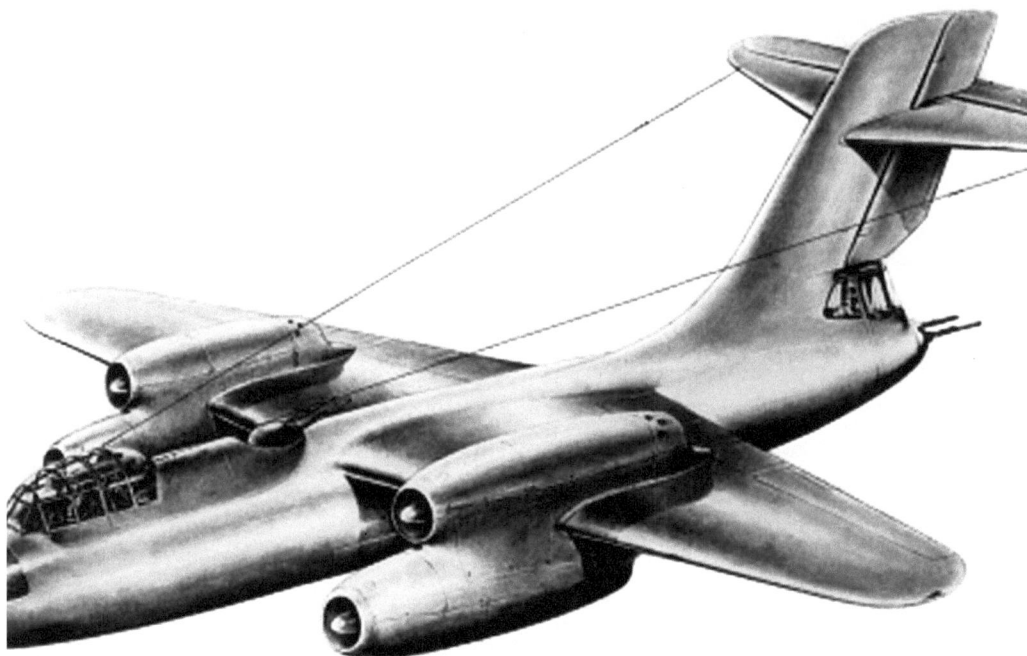

Some of Sukhoi's earliest designs were intended for the ground attack and bomber roles. Sukhoi also designed the Su-12 (top) as a bomber/trainer in the 1940's. As early as the second half of the 1940's Sukhoi had been involved in jet powered bomber design work, specifically the Su-10 (above), a four engine medium bomber of 1948 that carried the factory code 'E'. This design, however, did not progress to flight test.
Sukhoi

Among the factors leading to the selection of the Sukhoi bid was the high projected speed of 3200 km/h, which would bestow upon the Soviet Union a missile carrier aircraft all but immune to NATO then current and projected air defences. Development of the T-4 was officially authorised in a Soviet government resolution dated 3 December 1963 (often erroneously stated as April 1963). This allowed the design team, headed by Sukhoi designer general (chief designer) N.S. Chernyakov, to ramp up work on the project (Sukhoi). The T-4 was designed using the latest variants of the Ural domestic computer systems, possibly the Ural-2325. Ural computers were designed, developed and implemented into service from 1959-1972. The first generation Urals (tube type) and BESM-2m computers were superseded by more modern BESM-6 (operational in Sukhoi OKB from 1973) and M-222 type computers in the early 1970's. Such systems would probably have been incorporated into the T-4 design programs prior to cancellation (Sukhoi).

Further design studies and ground test experiments refined the road forward to achieve the operational requirement of cruise flight in the region of Mach 3 at altitudes in excess of 20000 m. To achieve this capability involved design collaboration with other institutions such as CAHI (TsAGI – Central Aerodynamic Institute), which 'produced comprehensive fundamental research into the aerodynamic performance of aircraft models [including wind-tunnel models], which made it possible to select the required configuration' (Sukhoi).

Following further design studies and ground test experiments, which included continued extensive testing of a number of model configurations in wind tunnels at TsAGI, the preliminary design review for the T-4 design concept was held in June 1964. The successful passing of this review later that year allowed progression towards a mock-up committee for the air forces of the USSR (Union of Soviet Socialist Republics). Following a further review in summer of 1964, the detailed design phase of the program was initiated under a joint team that included the MKB Burevestnik Engineering Design Bureau, and, in that year, the TMZ (Tushino Machine-Building Plant) was contracted to build the prototypes/development articles of the T-4 (article 100), the design still projected to be in the 100 ton class. Further refinements to the design continued in such areas as the arrangement of the engines as the program continued toward providing a close to final configuration for the mock-up committee for the air forces of the USSR, which was completed in February 1966. A team from Sukhoi Design Bureau had a meeting with the recently appointed Minister of Defence of the USSR, Marshal A.A. Grechko, in summer 1967. The end result of this meeting was a decision to accelerate the program. However, the approach of the leadership with the ministry of aircraft production was such that the contrary was in fact the case, work being considerably slowed (Sukhoi).

The selected design settled on what Sukhoi described as a 'flying wing concept', effectively a double delta (cranked) wing design, featuring a 'low margin of pitch stability and small size canards' that were employed in pitch when adjusting aircraft trim (Sukhoi). The double-delta wing itself featured 'a sharp leading edge and middle surface deformation' (Sukhoi). The forward fuselage had a diameter of 2 m.

The MiG-25 was designed as a ~Mach 3 interceptor as a counter to American aircraft capable of operating in the Mach 3 flight regime. The prototype flew in 1964 and the design was in widespread operational service with the Soviet air defence forces in the 1970's as the world's first operational interceptor aircraft designed for flight at ~3000 km/h.

The Mikoyan MiG-25 (NATO reporting name 'Foxbat') interceptor that conducted its maiden flight on 6 March 1964 had proven the Soviet ability to overcome the high temperatures encountered during aerodynamic flight at speeds in the region of Mach 3, as had the E-166 of 1961. This aircraft, which featured a take-off weight in the region of 41200 kg, was capable of flying at a speed of 3000 km/h with a ceiling of 23000 m. However, the MiG-25, like the American Lockheed SR-71 ~Mach 3 reconnaissance aircraft, an aircraft type that, among others, the MiG-25 had been designed to intercept, was designed to fly at ~Mach 3 for shorter periods compared with the sustained periods to be flown at such speeds by the T-4.

In order to meet the requirement for sustained flight at ~Mach 3 the normal use of aluminum alloy had to be replaced with other, more durable materials. The ability to withstand such high temperatures for extended periods of time was a major preoccupation of the Sukhoi design team (this included other bureau and organisations). To solve the problem much of the airframe structure was constructed of welded titanium alloys, steel alloys and small quantities of other material, allowing the aircraft to withstand the increased temperatures that would be encountered when flying in the region three times the speed of sound. Specifically the alloys included in the airframe construction apparently included the titanium alloys вт-20 (VT-20), вт-21л (VT-21L) and вт-22 (VT-22), stainless steel alloys VIS-2 and -5, steel вкс-10 (VKS-10) and maraging steel внс-2 (VNS-2) (this alloy, which contains up to 25% nickel and other metals, is strengthened through a process of heating, slow cooling and hardening). It was not, of course, simply a matter or developing materials for use in the airframe construction, a number of engineering solutions had to be pioneered for use in the that construction. Such processes included 'automatic through-penetration welding, automatic buried-arc welding using sheet add-on, chemical milling of titanium alloys, and others' (Sukhoi). The various new materials, coatings and manufacturing processes were extensively tested in a program that was a part of the overall T-4 design and development contract.

Top: In the second half of the 1960's TsAGI (CAHI) was involved in wind-tunnel testing of various aerodynamic models attributed to the SST program that spawned the Tu-144 and the ~Mach 3 strike platform that led to the T-4. TsAGI Top: The control panel laboratory for the high-altitude heat test stand (thermobarocamera) TBK-120 that was designed to test various equipment and components for the Sukhoi T-4 aircraft. Sukhoi

Trio of stills showing the T-4 during various phases of assembly in the assembly hall. The centre still shows the cockpit section with the pilot entry hatch in the open position while the bottom still shows the navigators entry hatch in the open position.
Sukhoi

Top: The first prototype of the T-4, article 101, during assembly in the assembly hall in 1971. Note the T-6 (Su-24) cockpit section adjacent to the T-4 forward fuselage. Above: Another image of T-4, 101, during final assembly – date unknown, but probably 1971. Sukhoi

The No.1 Convair XB-70A prototype lifts off on a development test flight in the 1960's. The T-4 would share many superficial similarities to the American aircraft, but both were fundamentally different in design and execution of mission. NMUSAF

At this juncture it is, perhaps, opportune to address the issue of the T-4 similarities, or lack thereof, to certain other designs of the period. To discuss the T-4 in the same context as that of the North American XB-70 is to distort the historical narrative. These were different species designed for different roles with a Mach 3 flight environment at high altitude in common. You can no sooner compare a Dolphin and a Human – both share some common environmental conditions and breathe air, but both are fundamentally different in design and function. The T-4 was designed for flight at marginally higher altitudes than that of the XB-70, both designs having projected maximum speeds in the region of Mach 3.1. However, although there were similarities in the performance regimes of speed and altitude, both designs were intended for fundamentally different roles – the former as a hypersonic missile carrier and the latter as a strategic nuclear bomber. This would lead to unmistakably different designs that adopted similar technology advances of their respective periods.

The XB-70, the prototype of which conducted its maiden flight in September 1964 (61-7976 achieved a speed of Mach 3 in October 1965), had a design maximum speed of Mach 3.1 (~3,300 km/h/~2,056 mph) at an altitude of 22250 m (73,000 ft.), service ceiling being 23576 m (77,350 ft.) and range was ~6900 km (4,288 miles). The range value for the American aircraft was slightly lower than that planned for the Soviet maritime hypersonic missile carrier. The XB-70 was considerably larger in overall dimensions other than height and had a take-off weight more than twice as high as the T-4 – loaded weight was 246618 kg (543,700 lb.), length 56.6 m (185 ft. 10 in) and height 9.7 m (30 ft. 9 in.). By contrast the planned Soviet maritime hypersonic missile carrier was designed with a take-off weight in the 102 ton = 92532.8 kg range and was planned to be considerably smaller than the XB-70 in overall dimensions other than height.

The increasing capabilities of advanced SAM (Surface to Air Missile) defences had led to the downfall of the XB-70, the projected capability growth of such defences put the huge bombers survivability in question by the time of its projected service entry or not long after. By contrast, the survivability of the T-4 concept was assured as it was designed from the start as an element of a weapon system that included the ~600 km range stand-off capability hypersonic aero-ballistic missiles, which would be immune to then current and projected near term defences. The stand-off capability of the concept vastly reduced the need for the T-4 to enter into an enemy's defensive zones when operating against ocean or coastal targets.

Analogies with civil aircraft designs like the Tupolev Tu-144/D and Anglo-French Concorde SST (Supersonic Transport)/passenger aircraft should, for the most part, be avoided. The Tu-144 prototype, 044, had conducted its 37 minute maiden flight from Zhukovsky on 31 December 1968 (pilots E. Eliyn and M. Kozlov) under the power of 4 x NK-144 turbojet engines, becoming the world's first supersonic airliner to fly (the Concorde prototype conducted its maiden flight just over two months later on 2 March 1969). The Tu-144D would, through a 1969 decision of the Committee of the Council of Ministers of the USSR, be powered by a derivative of the RD-36 (RD-36-51) that was specified to power the T-4 (RD-36-41), although the four engine arrangement was considerably different – the T-4 employed what was referred to as a single bathtub arrangement to house the engines on the aircraft underside whilst the Tu-144/D carried its engines in a two nacelle arrangement (this differed in detail in the two variants – Tu-144 and Tu-144D) on the aircraft underside. Although there are aerodynamic similarities, the design of the Tu-144 and T-4 were fundamentally different – the latter designed for sustained cruise at ~Mach 3 whilst the former was designed for sustained flight at the less demanding speed of ~Mach 2 (initially the Tu-144 upper speed limit was set at 2500 km/h) and maximum operating altitude, at 18000 m, was significantly lower than that specified for the T-4. The Tu-144 was also significantly heavier than the T-4 with a specified take-off weight of 207000 kg (~228 tons).

Tu-144 SST prototype with the nose section in the downward (drooped) position (top) and in the aerodynamic raised position (top). Tupolev

In terms of aerodynamics the T-4 had more in common with the Myasishchev M-53 supersonic strike/bomber concept of 1959 than it had with the XB-70. The M-53 concept had adopted a large double delta wing layout with forward canards, a single vertical tail and the four powerful engines located below and within the rear wing section, many of the traits observed on the later T-4.

T-4 nose section in the downward (drooped) position top and a port profile view of the T-4 prototype with the nose in the raised aerodynamic position (above). Sukhoi

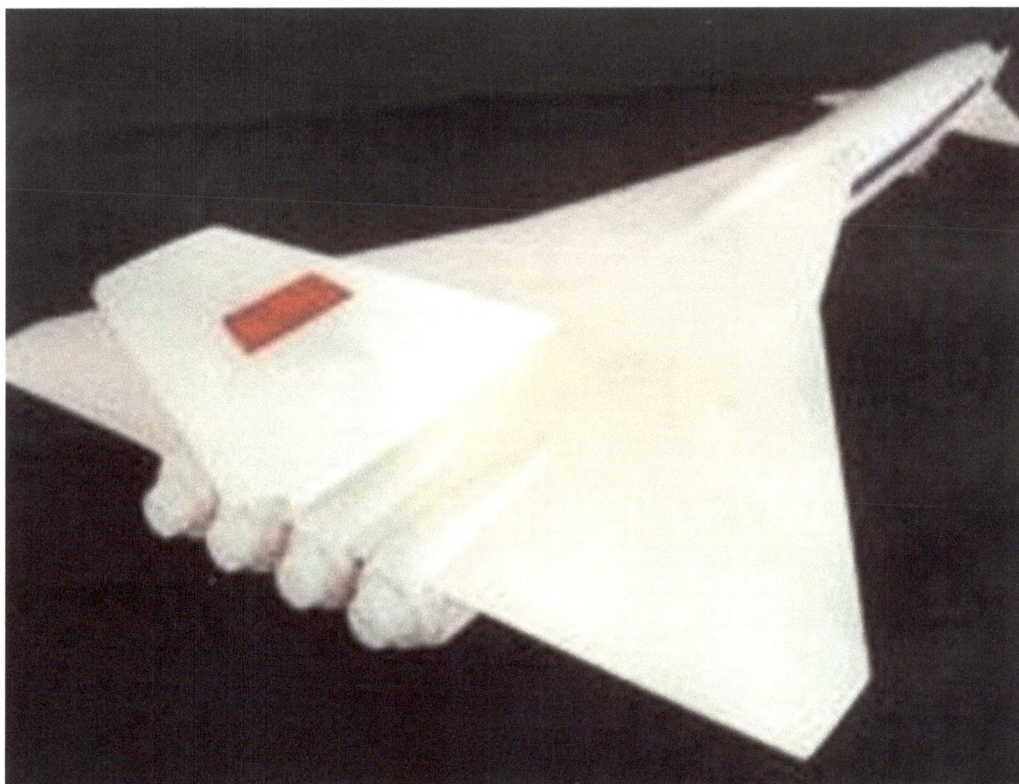

The T-4 shared a number of design traits with the appreciably larger Myasishchev M-53 concept of 1959 (above). Myasishchev

The major design trait that the T-4 shared with the Soviet and Anglo-French SST programs was the hinged nose concept, although in the SST's the pilots were still afforded a forward view when the nose section was moved upward to the optimum aerodynamic position. The T-4, unlike the civil SST programs, was flown completely on instruments during this phase of flight, which, in the T-4, was flown at considerably higher altitudes than the civil airliners.

The moving nose design characteristic, through automatically controlled hydraulics, allowed the nose of the aircraft to be rotated some 12° in yaw (upward/downward) when on the ground and in the take-off and landing phases and could be employed when operating in slow speed regimes under ~700 km/h. When operating above such speeds, at high altitudes (up to 22000-24000 m), the nose would be rotated upwards to reduce aerodynamic resistance of the forward area, including the cockpit section, which would now be more streamlined. When operating in this mode the pilot had no forward view, the aircraft being flown completely on instruments. When the aircraft descended to lower altitudes and speed reduced to 700 km/h, the nose would be rotated downward, affording the pilot a forward view of the outside environment. Such an innovative approach met with some disapproval, indeed, outright hostility by some senior defence officials. However, the concept won through with, at the suggestion of the designated chief test pilot on the project, Vladimir Ilyushin, provision for a periscope system that would afford the pilot a view of the aircraft outside environment in the event of an emergency such as a hydraulic failure that would prevent the nose section from rotating to the downward position for landing.

The nose section that would house much of the navigation/attack complex had to be designed to be tolerant to the high temperatures that would be generated at ~Mach 3 speeds, but still retain excellent radio-transparency to allow the function of the radar complex. The image at top shows T-4, 101, with leading test engineer on the program, A.S. Titov, beside in summer 1973. Above: The radio-transparent nose section of T-4, 101. Sukhoi/MODRF

One of the major challenges for the development team was the design of the nose fairing that had to withstand temperatures up to 4000° C, which could be generated during Mach 3 flight at high altitudes, whilst still retaining its radio-transparency to allow for the unimpeded function of onboard sensors such as the Navigation/Attack radar complex. To accomplish this the design team adopted a multi-layer (five) structure manufactured of honeycomb design material reinforced with glass fiber filler that would be impregnated with material tolerant to the high temperatures to be encountered. The thickest layer was the centre layer while the outer layer was coated with silicone for enhanced corrosion resistance.

Three view general arrangement drawing depicting the first prototype T-4, article 101. The illustration shows the general layout of the aircraft, the hinged nose section, cockpit sections(s), cockpit entry hatches, canard fore-planes, main wings, centroplan (centre wing section), elevons, vertical tail fin with two-piece rudder, 'bathtub' arrangement housing the four RD-36-41 afterburning turbojet engines with intake splitter section and tricycle undercarriage arrangement. Sukhoi

The T-4 main undercarriage consisted of two eight wheel main bogies (sometimes described as four wheel double tire, but, perhaps, more accurately eight wheel) that retracted into the fuselage underside, rotating 90° in the process. The twin wheel steerable front unit retracted aft to lay in the forward extension of the engine bathtub housing between the intake tunnels.

The undercarriage of the T-4 was designed along the tricycle principle with two main bogies, each of eight wheels (sometimes referred to as four double-wheels) and a twin-wheel steerable nose unit. MODRF

The major control surfaces consisted of three piece trailing-edge flaps (elevons) on each wing and a two piece hinged vertical rudder on the single vertical fin. The wing apparently had 0° anhedral and an angle of the order of 75° at the inboard leading edge, this adjusting to ~60° toward the outer span section. The major characteristic of the forward fuselage was the canard fore-planes which started just forward of the navigators cockpit entry hatch and extended rearward to a point just ahead of the start point of the double-delta wing sections. The canards, which adopted a quasi-lambda layout with a leading edge sweep in the order of 55°, were not designed as all-moving units as is the modern practice, but featured trailing edge flaps, which, as noted above, were employed in pitch when adjusting aircraft trim.

Previous page: The starboard main wing section (top), vertical tail, incorporating the two-piece rudder section, (centre) and the canard fore-plane section of T-4, 101. This page: Photo that purports (unconfirmed) to show some of the cockpit instrumentation representative of T-4, 101. MODFR/Sukhoi

The two crew entered the cockpit(s) section through entry hatches on the fuselage upper surface (roof). The layout was tandem with the pilot in front and the navigator in the aft section. Each of the hatches had small windows, this being the crews only outside view of the world around them once the hinged nose was in the (closed) aerodynamic position obscuring the forward view, the aircraft, as noted above, being flown purely on instruments at this stage. Escape was initiated by ejection of the crew K-36 ejection seats through the discarded entry hatches on the fuselage upper surface. The crew would wear high-altitude pressure suits and were supported by an internal life-support system that included oxygen generation and environmental control. Aft of the crew compartment was a cooled bay for electronics.

Just as the airframe concept was refined a number of times so too was the choice of engine to power what would emerge as the T-4 design. Early studies of R-15B-300 (this was the engine that powered the MiG-25 ~Mach 3 capable interceptor) and RD-17-15 engines were quickly dropped as these were deemed unsuitable for the power requirements at the high operational altitudes that were specified. The solution to this problem was an adaption of the RD-36 series then under development by Rybinsk Engine Building Design Bureau (OKB-36), now NPO Saturn (a subsidiary of United Engines Corporation). The RD-36-41 was selected to power the T-4 following detailed studies on the power requirements and options available in the required timeframe.

Graphic of the Rybinsk (NPO Saturn) RD-36-51 that powered the Tu-144D SST. This design had much in common with the RD-36-41 from which it was developed. UEC

The RD-36 series could trace ancestry back to the VD-7 series. These powerful engines, development of which had commenced in 1954, had powered a diversity of aircraft types, including the Tupolev Tu-22/K and the Myasishchev M-50 supersonic missile carrier/bombers (Saturn). In 1961, P.A. Kolesov took over as chief designer at the Rybinsk Engine Building Design Bureau and led the development effort for the RD-36 series of afterburning engines (Saturn & Sukhoi). Kolesov had himself been the lead in the design of the VD-7B gas-turbine that powered the M-50 and headed the bureau when it was contracted (in 1960) to provide the VD-7M-2 afterburning engine to power the Tu-22K supersonic missile carrier (Saturn).

Development of the RD-36-35 had commenced in 1963 to power various tactical combat aircraft, including the MiG-21PD, MiG-23PD and Sukhoi T-58VD, the RD-36-35VFR lift engines for the Yak-38 (Yak-36M) V/STOL (Vertical/Short Take-Off and Landing) fighter and the RD-36-51K boost engine for the EPOS aircraft element of the Spiral spaceplane program. The RD-36 would be developed into the significantly more powerful RD-36-41 to power the T-4. This engine, which incorporated a single-shaft eleven-speed compressor section, a two-stage turbine section equipped with air-cooled turbine blades and an afterburner section, was the first supersonic capable mixed-compression engine fed air through variable air intakes that had thus far been developed in the Soviet Union. The design featured an auto-start system and was designed for extended period cruise operation at a speed of ~Mach 3 (Sukhoi). The RD-36-41 featured a rudimentary automatic control system, more advanced systems of which are standard in modern engine designs. The exhaust nozzle itself featured a three stage afterburner section.

Through several design iterations it was decided to house the engines in a large nacelles on the aircraft underside, this, as noted above, being referred to as the 'bathtub' configuration. For testing of the RD-36-41 a power unit test stand, which incorporated a T-4 aircraft engine air intake was built. This stand, designated RMCS, was used to ground test the RD-36-41 at Turaevo from 1970 (Sukhoi).

The RMCS test stand for testing the RD-36-41 at Turaevo circa 1970. Sukhoi

Development of the RD-36-41 took ~ten years, the basic design spawning a number of derivatives, being the bedrock for design of the RD-36-51A to power the Tu-144D Mach 2 capable SST and the RD-36-51V, which was selected as the power plant for the Myasishchev M-17 high altitude-aircraft designed to conduct research in the stratosphere (Saturn).

The fuel for the RD-36-41 engines was housed within three T-4 fuselage fuel tanks, a forward tank, the main (central) tank and an aft tank. These held a combined total fuel load in the region of 57 tons (this value may be adjusted slightly upward or downward). Various figures have been circulated for fuel consumption rate, it being sound to provide a mean in the order of 1.8-9 kg/kg/h when afterburner was used. There was, of course, such items of equipment as systems for emergency jettison of fuel.

Top: The engines were arranged four abreast with the afterburner section/exhaust short of the wing trailing surfaces. The protruding tail section accommodated the fuel dump system and the housing for the brake parachute. Centre and above: The T-4 inlet arrangement with splitter plate extension (centre). MODRF

Others systems installed in the T-4 design included a 'trailblazing fuel system with turbine driven pumps', an auto-throttle, a hydraulic system specified with an operating pressure of 280 kg/cm^2, a 'liquid-nitrogen-based inert gas system' as well as many other systems (Sukhoi).

An extensive multi-bureau research effort was undertaken to design the many systems required for the T-4's demanding flight environment. Not least among these systems was the design and development of what was then referred to as the SDU (Remote Control System) now referred to as a FBW (Fly-By-Wire) FCS (Flight Control System). This innovative system would oversee the various flight control units of the aircraft. The T-4 was the first Soviet designed aircraft to be equipped with such a quadruple redundancy FBW FCS and would, on the event of its maiden flight, have the distinction of becoming the first aircraft of any country to fly under the control of such a system. This first generation FBW FCS operated in the Z-M channels and featured quadruple redundancy with a provision for reversion to manual control in the event of the failure of two or more channels.

In the cockpit the crew of two operated the advanced (for the time) instrumentation, navigation and targeting systems and communications complex. The heart of the planned operational mission equipment suite was the Vukhr fire control system and the NK-4 navigation system, but the suite also incorporated a number of other systems, many integrated within the Soviet Okean avionics suite. The navigation/attack suite would have been capable of comparing generated data with a detailed library of vast portions of the territory of the planet, including, of course, oceanic areas in line with the primary role – data from which would be displayed on a screen(s) in the cockpit(s).

Self-defence was provided by the Otpor defence system and the major communications means was conducted through the Stemnina radio communications complex (Sukhoi).

Depiction of the F-020 reconnaissance aircraft concept of 1957. Myasishchev

To say that the Soviet Union was, in the late 1950's and into the 1960's, perplexed at NATO, predominantly United States, aircraft violations of its airspace on reconnaissance and nuisance overflights may be a colossal understatement. While the Soviet Union actively countered the overflights with a number of high profile shoot downs of NATO reconnaissance aircraft, it refrained from embarking on an active overflight program of its own. However, studies were conducted on the viability of developing a survivable inhabited aircraft reconnaissance platform. Among the many concepts arrived at was the NB-1 and Myasishchev F-020 supersonic reconnaissance aircraft of 1957. Going into the 1960's, the potential posed by the introduction of reconnaissance satellites largely negated the need for an active overflight program. The first Soviet reconnaissance satellite, the Zenit-2 (Kosmos-4), was launched into orbit aboard a Vostok-K launch vehicle from Tyuratam, Baikonur, cosmodrome in Soviet Kazakhstan on 26 April 1962 (MODRF). The vehicle orbited Earth with a periapsis of 285 km and an apoapsis of 317 km for some three days before the reentry capsule was returned to Earth on 29 April 1962 with a cargo of photographic images of sites of interest to the Soviet high command (the role of this satellite is erroneously listed by NASA as being to study the upper layers of Earth's atmosphere and near space environment, but has been confirmed by the MODRF as being a photographic reconnaissance mission). In this atmosphere of a viable survivable strategic reconnaissance capability from Earth orbit it can, with confidence, be assumed that the reconnaissance mission that was to be inherent within the T-4 concept was aimed at providing a survivable oceanic reconnaissance capability in order to locate targets for missile carrier T-4 aircraft or other maritime strike platforms, including surface warships and submarines armed with ASCM. That said, the reconnaissance capability would probably have been such that it could have been called upon to cover inland target areas in the absence of imagery from reconnaissance satellites.

For the reconnaissance role the production variant of the T-4 was to be equipped with a Rapira reconnaissance system for which practically no hard facts are available. It is equally unclear if a specialist reconnaissance variant would have been developed or if a lower-scale reconnaissance capability, built into the main strike variant, would have met the reconnaissance requirement. Despite the aircraft's Mach 3 capability, in the absence of a Soviet Union or Eastern Bloc overflight mission for the USAF (United States Air Force) SR-71 Blackbird ~Mach 3 reconnaissance aircraft, it is doubtful if a production T-4 variant would have been employed on reconnaissance overflights of the western hemisphere. Such a requirement had, notwithstanding the fact that the Soviet Union did not previously have an operational policy of deep overflights of NATO territory that contravened international law, as noted above, been largely negated by the increasing capability and availability of reconnaissance satellites for both power blocks of East and West.

As well as having the ability to locate targets independently, an operational T-4 variant would have been allocated targeting information from other airborne assets such as Tupolev Tu-142 long range maritime reconnaissance aircraft, Ilyushin Il-38 (NATO reporting name 'May') long range maritime patrol aircraft and, as noted above, other T-4 aircraft performing a reconnaissance target acquisition mission. In addition, the T-4 would have incorporated an equipment suite that enabled the aircraft to receive targeting data from other platforms such as submarines, surface warship/surveillance ships and information originating from the nuclear powered RORSAT (Radar Ocean Reconnaissance Satellite), which was designed to search vast ocean areas for NATO shipping from a low-Earth orbit (the first RORSAT satellite, Cosmos 209, was launched atop a Cyclone-2 launch vehicle on 22 March 1968).

As Mach (the speed of sound is not a constant, but rather depends on the variable of altitude, due to air temperature and density, it is best, for the most part, to refer to the projected maximum speed of the T-4 in km/h. The basic performance parameters included a cruise speed of 3200 km at high altitude with a practical operating altitude of 24000 m. At 18288 m altitude this would equate to a speed of ~Mach 3.01. Maximum speed of the T-4 at sea level was in the region of 1100 km/h. A production T-4 variant would have been unlikely to operate at sea level as the aircraft design, like the X-45 hypersonic missile armament, were optimised for multi-Mach flight at altitudes of 20000+ m. On paper it was expected that a T-4M derivative could achieve 3500 km/h. If this was attained at 18288 m altitude then a speed of ~Mach 3.295 could have been achieved. The T-4 practical range of 7500 km would have allowed targets to be engaged at considerable distances – perhaps ~3000 km from launch base, not taking the additional range of say several hundred km afforded by the X-45 missile after release from the launch aircraft, assuming an active homing capability. This range could have been extended courtesy of an inflight refueling system that was planned for series production aircraft, but, not installed on the T-4 prototype. Take-off run was ~1000 m while the landing speed was in the region of 260 km/h with a landing roll of ~950 m.

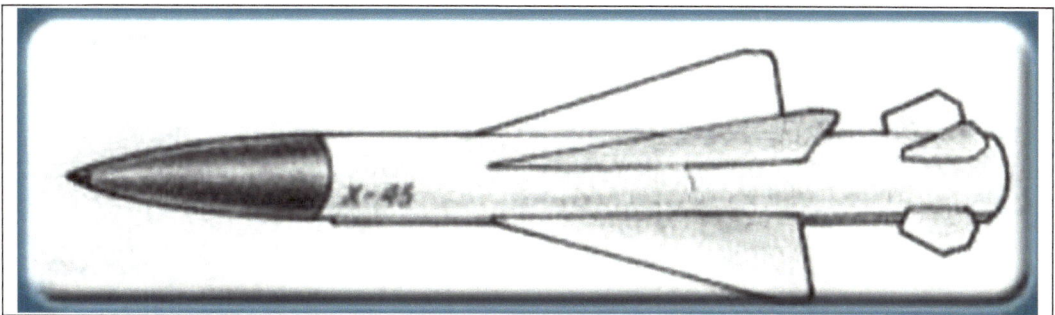

Graphic depicting the MKB Raduga X-45 (Kh-45) hypersonic aero-ballistic missile.
DMB

X-45 HYPERSONIC AERO-BALLISTIC CRUISE MISSILE – The sole armament specified for the T-4 was two x MKB Raduga X-45 (Kh-45) hypersonic aero-ballistic cruise missiles. This advanced missile, which was being developed simultaneously with the T-4, had a projected flight range of 550-600 km and was designed to cruise to the target at a speed of between Mach 5 and Mach 7 (Sukhoi).

Although design of the Kh-45 had been initiated at Sukhoi, design work was transferred, at an early stage in development, to Raduga. The General Designer on the X-45 program, A. Ya. Bereznyak, had been a leading figure in Soviet cruise missile design from the early post war years. This included involvement in the KS air launched ASCM (1951) and C-2 KSS (1954) land launched ASCM (general designer on these programs was A.I. Mikoyan of O.K.B. Mikoyan). Bereznyak went on to become general designer on the CS-7 mobile ground launched cruise missile (1954) and K-10S air launched ASCM intended for carriage on the Tu-16K-10 and Tu-16-10-26 medium maritime missile carrier/bomber aircraft (1955). Under the administration of ICB 'Rainbow', Bereznyak oversaw the design of the sea platform launched P-15 ASCM (1956), the KSR-2 and KSR-11 (1956) ALCM to arm the Tu-16-11-16, the X-22/M/MA/N/HA (1960) ALCM to arm Tu-22K/Tu-22M2/M3 and Tu-95K-22 bombers, KSR-5/P air launched ASCM (1963) to arm Tu-16K-26/Tu-16-10-26, the X-28 anti-radar missile (1963) for tactical strike aircraft of the Su-17M/M2/M3 and Su-24 types and the 85R (1967) and 85RU(PM) (1974) ship launched anti-submarine cruise missiles.

Development of the X-45 was initiated with the aim of arriving at a hypersonic speed high precision missile capable of striking mobile naval targets such as an aircraft carrier, as well as fixed location targets such as ports. Such a weapon, like the carrier aircraft, had to be designed for flight operations at very high temperatures. In this regard, the X-45 program should not be considered simply as an anecdote to the T-4 (and the later T-4M) portrait, but rather should be considered as an integral part of the Sotka weapon system. Development of this missile was fraught with technological difficulties that were on a par with those posed for the design of the carrier aircraft.

It is a point of fact that development of air launched hypersonic missiles has proven to be one of the most technologically challenging areas of aeronautics over the four plus decades since the cancellation of the T-4/X-45 complex. Only in late 2017 was the first truly hypersonic air launched missile system introduced to service in the shape of the Kinzhal Air System – consisting of a MiG-31K Mach 2.83 capable launch platform (originally designed as an interceptor) carrying a single hypersonic missile, for much the same role as that planned some four and a half decades earlier for the T-4/X-45. The survivability benefits of hypersonic flight speeds came at a price, the missile airframe being subjected to a considerably higher level of stresses and temperatures than a missile designed for subsonic or even high supersonic speeds. The hypersonic flight speeds that would be encountered dictated much of the materials chosen for the X-45 body construction. The main materials consisted of 'stainless steel EI-654 (tanks); titanium alloy VT-20 (compartments GCHF, HCHF, wing, tail); Separate power elements were made of high-strength steels 12XTBBφA, 30XrCA, steel BHC-3, titanium alloys OT4-1, OT-4, BT-5, BT-5-1, BT5-ʌ and magnesium alloy Mʌ-10' (DMB). At the design development phase various solutions were investigated in order to reduce the overall complexity of the weapon complex. Among such innovations arrived at was 'a steering frame with a rim made of VT-5L alloy and a cross from alloy ML-10 (A.Z. Chuchalov, Yu. F. Chetverikov, M.A. Borisova, A.A. Bykov). Removable panels of heat protection of

OT-4 and basalt fiber Type B (A.A. Osochenko); tunnel tubes for laying communications in tanks (V.M. Kulikovsky, M.Z. Eidlin, O.V. Melnikov)'; a system to allow for the 'quick-detachable fastening of the front fairing with the cargo [payload] compartment with the help of a clamp (O.V. Melnikov, G.F. Chetverikov)' (DMB). One of the most challenging problems faced by the design team was in developing the radio transparent fairing, which had to be corrosion resistant and heat resistant to withstand the very high temperatures generated at hypersonic speeds. To overcome this problem a multi bureau team was assembled with specialists from DMB, VIAM and VNIISPV (designers G.A.A. Dmitrenko, S.A. Kumankova, V.V. Skortsov, L.E. Kurilov, I.V. Pugacheva, G.A. Osilkin, E.S. Panov, V.S. Zimonkov, V.V. Pavlov, V.N. Potashova, V.N. Bulgakov, R.V. Raikov and N.V. Nozhukhova). Through extensive experimental testing the design arrived at was a 'two-layer monolithic structure', which was then extensively tested, including tests of the quick-detachable fastening system mentioned above (DMB). The fairing structure was constructed from 'a power layer of SK-9FAK glass-textolite on the basis of a quartz fabric and silicone-bonding agent K-9FA; an erosion resistant layer of SK-4K glass-textile on the basis of a multi-layer all-rolled quartz cover (specially developed by VINIISPV) and an organosilicon K-4 binder' (DMB). The manufacture of the materials, as stated by DMB, was conducted through a process of 'impregnation under pressure (with back pressure)' (DMB).

Providing the missile with hypersonic flight speeds was only one of the major challenges facing the design team. The missile operational doctrine called for very high accuracy against moving targets, no easy feat in the 1960/1970's. To enable this the design was to incorporate an advanced inertial navigation and control system 'based on giroinertsialnoy system' the functions overseen by a digital computer (DMB). The onboard radar for target acquisition and homing is often speculated to be of the semi-active radar hoping type, however, the autonomous operation of the weapon would strongly suggest an active homing capability was incorporated, this negating the need for continuous target illumination by the launch vehicle or other targeting source.

The X-45 layout adopted was that of an aerodynamic body with an 'X' or cruciform main wing layout aft of centre and an 'X' or cruciform smaller control surfaces at the rear, just forward of the missile exhaust section. The power requirement for hypersonic speeds with an aero-ballistic capability was to be met with one single-chamber rocket engine. This ensured a high degree of maneuverability was available for the missile flight profile. Initially the X-45 had a projected weight of ~4500 kg. The warhead(s) design had not been finalised at the time of cancellation, but were expected to be of two types, a 500 kg class conventional warhead for the anti-ship strike mission and a nuclear warhead of undetermined mass for the nuclear strike mission against mobile maritime or fixed land targets. This latter warhead would have had a similar yield to the 10 kiloton warhead of the K-10 ASCM, which was tested at the Novaya test site on 22 August 1962 following launch from a Tu-16K some 400 km from the impact point (MODRF).

As noted above, operation of the X-45 was to be fully autonomous. After launch at an altitude of ~20 km the missile would very quickly accelerate to hypersonic speed and cruise at such to the target area receiving navigational updates through the onboard navigation complex. The trajectory chosen for the missile was the most economical, in regards to required energy expenditure. A number of flight modes would have been available, allowing a diversity of attack profiles to be incorporated into a number of potential mission scenarios against different target sets. Nearing the target area the missile onboard radar complex would locate the target, the complex computer system processing the target information to calculate the most appropriate terminal phase attack profile.

Two poor quality stills showing the T-4 prototype, 101, taking off from Zhukovsky on its maiden flight on 22 August 1972. Sukhoi

Construction of the first T-4, article 101, which had commenced in 1966, was completed in the autumn of 1971. After some initial ground testing at the factory the aircraft was, in December 1971, transferred to the FRI airfield, Zhukovsky, in preparation for its maiden flight, which, following several months of delays due to adverse environmental conditions, was conducted on 22 August 1972. This, as noted above, was the first time that an aircraft had flown under the control of a FBW FCS. Crew for this historic flight was V.S. Ilyushin (pilot) and N.A. Alforov (navigator).

The same crew would fly the aircraft for all ten of its flights, the last of which was completed on 19 January 1974, by which time the aircraft had attained a speed of Mach 1.36 at an altitude of 12000 m (Sukhoi).

Top: The prototype T-4, article 101, during flight testing with the undercarriage extended (the most reliable evidence suggests that this was during the aircraft's maiden flight on 22 August 1972). Above: The crew of V.S. Ilyushin (pilot) and N.A. Alforov (navigator) exiting the aircraft following the successful completion of the maiden flight of T-4 article 101. Sukhoi

Top: The prototype T-4, 101, takes-off from Zhukovsky under the combined power of the 4 x RD-36-41 afterburning turbojet engines, each of which generated ~16000 kg thrust. Centre and above: T-4 101 lands at Zhukovsky, the landing run being reduced through the deployment of a quadruple canopy brake parachute system. *Sukhoi*

Trio of images showing T-4 101 during flight testing. Sukhoi

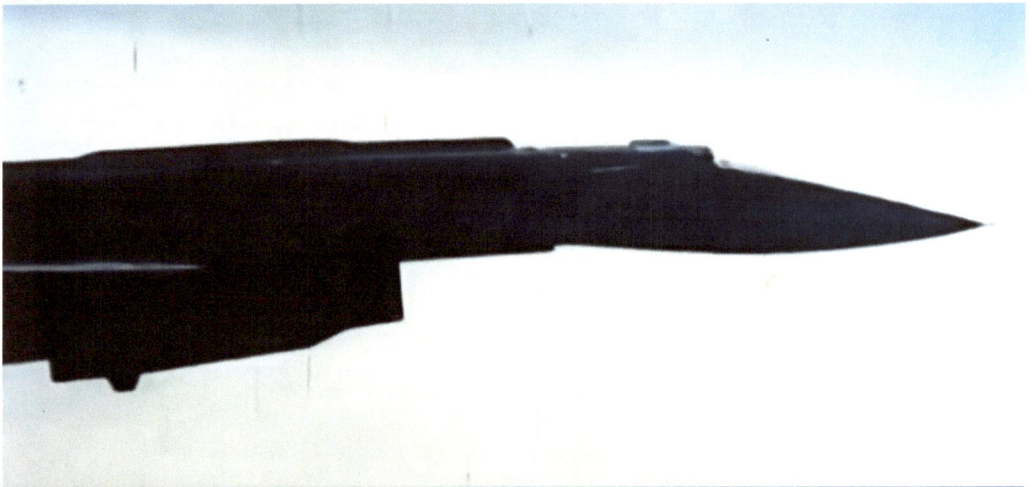

The images on this page show the T-4 prototype with the moveable nose section in the downward (drooped) position (top), during the transition to the aerodynamic position (centre) and in the aerodynamic position for high speed (multi-Mach) flight (above). Sukhoi

A total of nine flights were conducted during the first stage of flight testing – the first five of which were flown at low speeds and with the undercarriage extended. The aircraft apparently attained Mach 1.28 on flight nine, which was apparently flown on 8 August 1973. The first flight of the planned series of flights under the second phase of flight testing was conducted on 22 January 1974, this being the last flight of the T-4 prototype. It was during the course of this flight that the aircraft attained an altitude of 12 km and a speed of Mach 1.36, the highest altitude and speed attained before flight testing was suspended pending a review on the future of the program. During phase two flight testing it had been planned to push the speed through the 3000 km/h milestone in preparation for extending to the maximum specified 3200 km/h. The intention was to commence flight testing on the second T-4, article 102, during phase two, but such plans were cancelled when the program was suspended. With the completion of the single phase two flight the program had accumulated a total of ten flights in a calculated flight time of ~10 hours and 20 minutes. Although there were undoubtedly issues to be addressed, for the most part the basic T-4 flight characteristics, in the flight regimes investigated, had proved to be sound.

Although article 101 was the only T-4 to fly, a total of four airframes were constructed or partially constructed – three flight test aircraft, article 101, 102 (mentioned above) and 103, and a static ground test example, the latter designated article 100S. A further three flight test aircraft, articles 104, 105 and 106, were in various stages of assembly when work on the program ceased following a USSR MAI (Minister of Aviation Industry) order for the T-4 program to be suspended in 1974. The program was officially cancelled by a resolution of the Soviet government issued on 19 December 1975 (Sukhoi). At the time of cancellation article 101 had been grounded in flyable condition, article 102 was more or less complete, but had not flown, and article 103 was partially assembled. Parts and materials existed for other aircraft, but these had not progressed to final assembly. The incomplete T-4 airframes were scrapped, although parts of article 102 were, for a time, stored in an aircraft hangar at the MAI, but these were eventually scrapped as was the partially assembled article 103 and components for the other planned aircraft. Article 101 was stored before being transferred to the Soviet (later Russian) Air Force Museum at Monino where it resides in the exhibit park in 2018.

Following the T-4 program cancellation Soviet air forces equipment planning put increased reliance on the Tupolev Tu-22M intermediate range missile carrier/bomber/reconnaissance aircraft design. The prototype Tu-22M0 had conducted its maiden flight on 30 August 1969 (pilot Borisov), the successful flight testing of this design, which entered series production in 1971, providing a cushion to fall onto in the event of T-4 cancellation. In the anti-ship role this Mach 2 capable variable-geometry (swing-wing) aircraft would be armed with the high supersonic Kh-22M series (NATO reporting prefix AS-4) air to surface missile providing a highly credible, but less potent, counter to NATO aircraft carrier battle groups threatening the Soviet Union in the last decade and a half of the Cold War – the Soviet Union was dissolved on 25 December 1991. The Tu-22M2/3 served with the major post-Soviet successor state, the Russian Federation, 165 and 100 serving with Russian Naval Aviation and Long Range Aviation respectively in late 1993.

Long before the resolution of the Soviet government ordering cancellation of the T-4 program Sukhoi had embarked upon studies of a variable-geometry (swing-wing) variant designated T-4M, concept development work being conducted in the period 1968-1970. During the period 1970-1972 the design was further evolved into the T-4MS, which received the code article 200, weight, in the ~200 ton class, increasing considerably over that of the T-4. This latter design had emerged as part of the competition in which other design houses, notably Myasishchev and Tupolev, were involved. The prize was a contract to build a strategic duel role aircraft for Soviet Long Range Aviation to meet a requirement for a high speed long range bomber to replace existing subsonic penetration strategic bombers like the M-4/3M. This competition was won by a Tupolev design that would eventually emerge as the Mach 2 capable Tu-160 variable-geometry long range strategic missile carrier. It was indicated that the T-4's demise was to allow for available funding to be allocated to the planned strategic missile carrier program, a role for which the T-4 was wholly unsuited. The traits of meeting the 100 ton requirement that was, in part, the reason for Sukhoi winning the competition to develop the maritime hypersonic missile carrier being a not insignificant reason behind the program cancellation. At the time of its cancellation it was decreed that some 1.3 billion Rubles had been spent on the T-4 project.

T-4 101 in flight with the undercarriage retracted (top) and extended (above). Sukhoi

Notwithstanding the events noted above, a number of additional narratives have been forwarded as to the reasons for the T-4 programs cancellation – funding being required for the large production program of Mikoyan MiG-23 variable-geometry tactical fighter aircraft being but one. This, however, is highly unlikely as both programs were initiated for very different operational environments. The choice of a large-scale tactical fighter program or a significant production run of maritime hypersonic cruise missile carriers in the class of the T-4 would never have been presented as an either or. The instigation of this particular theory probably stems from the fact that as TMZ (Tushinsky Machine Building Plant) was not in a position to enter large-scale production of the T-4 it left the door open for manufacture of the wing sections for the MiG-23 at the plant. Planned production of the T-4 had been expected to stretch to at around 250 units. While it was TMZ that was contracted to build the static and flight test T-4 development aircraft, series production would have been undertaken at the Kazan plant (now Kazan Aviation Production Association), this being a cause of concern for Tupolev, which relied on the plant's capacity to produce Tupolev aircraft – a shortfall in building capacity may well have ended Tupolev's hopes of continuing with missile carrier/bomber projects such as the Tu-22M series and the later Tu-160.

Ultimately, despite losing the competition that spawned the T-4, Tupolev would retain its position as the Soviet Union's premier long-range missile carrier/bomber designer, not only advancing with the Tu-160, but also embarking on a large scale production effort for the T-4's ersatz replacement program – the Tu-22M series (the T-4 would have been vastly more expensive to procure than the Tu-22M series (this ultimately contributed to the programs demise), which in Tu-22M3 (some Tu-22M3's are being updated to Tu-22M3M standard) guise remains in service with the Russian Federation in 2018, as does the Tu-160 (in the second half of the second decade of the twenty first century production of the Tu-160 recommenced to build a fleet of Tu-160M strategic missile carriers for service in the 2020's and beyond), the latter aircraft, armed with subsonic cruise missiles of the Kh-55 (nuclear warhead), Kh-101 (conventional warhead) and Kh-102 (projected nuclear warhead) types. Despite being inferior in many performance parameters to the T-4/M, development and series production of the Tu-22M series was, as noted above, possible at vastly reduced cost in comparison to that of the Sukhoi aircraft, allowing funding to be made available for the Tu-160 program. While this latter aircraft would carry the X-55 nuclear armed strategic ALCM, the Tu-22M series would carry the high supersonic X-15 (1978 – general designer I.S. Seleznev) and the longer range high supersonic X-22M series ALCM. In this respect, just as the Tu-22M series had taken the place of the T-4, the X-22M took the place of the X-45.

Despite its cancellation as a weapon system, the T-4/X-45 programs had resulted in a major advancement of the Soviet Union's aviation and missile industry. In regards to the aircraft itself, the airborne systems and assembly process, 208 new inventions had to be developed and implemented into the overall program. When the various assembly components and other units incorporated within the program are taken into account, the number of inventions resulting from the T-4 design and development program is extended out to ~600.

Months after this photograph of the T-4 prototype, article 101, was taken in summer 1973, the writing was effectively on the wall as the Soviet defense expenditure purse was overhauled. NPO Saturn

Although the T-4 program had fallen by the wayside, the threat it was designed to counter was still very much in existence. This, as noted above, left the door open for a cheaper, technologically easier to attain solution to counter the aircraft carrier battle group – Tupolev being the obvious choice to fill the requirement with its Tu-22M project armed with X-22. Had the T-4 program been carried through to fruition, it would have been at the expense of the Tu-22M variants, although it remains a point of contention as to whether or not the Tu-22M series would still have been built for Long Range Aviation or if that service would have adopted a variant of the T-4 for the nuclear armed cruise missile strike and bomber roles. The Mach 2/high supersonic capable Tu-22M/Kh-22 combination bestowed upon the Soviet air forces a highly capable intermediate range strike/anti-ship strike capability, albeit with considerably less survivability to that offered by the Mach 3/Mach 7 capable T-4/Kh-45 combination. However, by studying the role of the Tu-22M series in Soviet naval aviation we get an insight into the probable operational role envisioned for the T-4, albeit both designs would adopt appreciably different flight profiles.

Tu-22M2 (NATO reporting name 'Backfire' B) over an oceanic area in company with an USN F-14A Tomcat fleet air defence fighter. USN

As the 1980's dawned the increasing capabilities of Soviet SSBN's dictated that a sizeable element of this fleet would be able to operate in waters nearer to the Soviet homeland than had hitherto been the case. This would allow the Soviets to more or less combine the requirements to protect the SSBN fleet and counter NATO aircraft carrier battle groups as both required access denial of considerable oceanic areas approaching the Soviet homeland. Defence of the Soviet SSBN force was practiced by an echeloned defense in depth principle for which conventional and nuclear powered cruise missile and or torpedo armed general purpose submarines, cruise missile armed surface combatants and land based long range ASW/patrol and strike aircraft would be employed. In addition, it was expected that Soviet SSBN's would be supported by Soviet SSN's during transit to their operating areas, some SSN's remaining in the general operating areas to protect the SSBN's against NATO SSN's. By the early 1980's, the operating areas for the second and third generation Soviet SSBN's would cover large areas of the Kara, Barents, Northern Norwegian and Greenland Sea's in regards to the western hemisphere. In the eastern hemisphere, the main SSBN operating areas were expected to be the Sea of Japan, Sea of Okhotsk and the general sea area of the Kamchatka Peninsula.

The Soviet strategy called for general sea access denial out to distances of some 2000 km from the SSBN operating areas. This allowed the SSBN defence policy to be integrated with the similar defence in depth policy applied to the defense of the Soviet Union from aircraft carrier battle groups. The most important forces for sea access denial would have been the cruise missile armed elements of the general purpose submarine fleets and the land based naval aviation anti-ship strike units, in particular those equipped with the Tu-22M2/3 intermediate missile carrier (introduced from 1974) armed with X-22M2 long-range supersonic air-to-surface missiles. These assets were to deny NATO aircraft carrier battle groups from operating in areas that would allow them to threaten Soviet territory or the innermost SSBN operating areas. The Soviets would have expected the operating areas for NATO attack aircraft carrier battle groups to include the Norwegian Sea, North Sea, Eastern Mediterranean Sea and the northwest Pacific Ocean.

From its introduction to Soviet Naval Aviation in the mid-1970's, the Tu-22M took on a primary role of sea denial armed with Kh-22M/2 long-range anti-ship cruise missiles designed for attacks on high value assets such as aircraft carriers. US DoD

Four Regiments of Tu-22M series aircraft were available for Soviet Naval Aviation by the end of 1982 – two in the Baltic Fleet, one in the Black Sea Fleet and one in the Pacific Fleet. The Baltic Fleet Regiments often deployed to Northern Fleet air bases during training exercises where they could provide cover for the Northern Fleet surface combatants, including the Project 1143 Kiev Class Heavy Aircraft Carrying Cruisers, employed on ASW operations in line with the primary role of protecting Soviet SSBN's. In short, a vast portion of the Northern, Pacific and Black Sea Fleet surface combatants, general purpose submarine and airborne assets would be employed on SSBN protection and or anti-aircraft carrier battle group operations, although for many deployed units the primary mission would overlap with other missions such as engaging NATO units not necessarily tasked with countering, or indeed being considered a threat to the deployed Soviet SSBN forces or the Soviet homeland.

While NATO assessments were clear on the threat posed by the formidable Soviet ASCM forces to its aircraft carrier battle groups and other naval assets, it can, with hindsight, be stated that the intelligence agencies erroneously assumed that the alliances capabilities to combat cruise missile attacks were greater than they in actuality were – assessing that the Soviets could achieve the destruction of carrier battle groups, but at significant loss in aircraft and missiles. The Anglo-Argentine Falklands war of April-June 1982 showed that modern warships were extremely vulnerable to even subsonic cruise missiles – Britain suffering the loss of a number of ships, including the Type 42 Air Warfare Destroyer HMS *Sheffield*, to subsonic speed air launched AM39 Exocet ASCM. This hard learned lesson was once again hammered home in 1988 when the United States Guided Missile Frigate USS *Knox* was heavily damaged by an AM39 launched from an Iraqi strike fighter. It is now clear than the Soviet launch platforms of the late 1950's through much of the 1970's were indeed vulnerable to NATO fighter forces, including carrier launched fighter aircraft. However, with the introduction of the Tu-22M series armed with high supersonic ASCM's, this vulnerability was considerably reduced, the ASCM armed Tu-22M3 considered to still present a considerable threat to aircraft carrier battle groups in the twenty first century.

In 2018, the Russian Federation operates a force of Tu-22M3's in the missile carrier/bomber roles and expected to receive the first updated Tu-22M3M before the end of the year. This latter variant will apparently be armed with a new Mach 5 speed cruise missile designated Kh-32. UAC

Following the cancellation of the T-4 program, the USSR continued development work on aircraft capable of flying at speeds approaching Mach 3. This was primarily aimed at improving the Soviet Union's air defence capabilities, resulting in the MiG-31 interceptor (NATO reporting name 'Foxhound'), the prototype of which conducted its maiden flight on 16 September 1975. Being designed as an interceptor advancement on the MiG-25, the MiG-31 was overall dimensionally smaller than the T-4 and had a lower overall performance in regards to maximum speed and altitude – ~3000 km/h and 20600 m respectively. While the MiG-31, in interceptor form, would serve the in the air defence role with Soviet and later Russian Federation and Kazakhstan air forces, it would, in the second half of the second decade of the twenty first century, take-on a limited maritime and land attack strike role armed with a single Mach 10 capable ~2000 km range Kinzhal Air System air launched aero-ballistic missile system (apparently designated Kh-47M2, but unconfirmed as of March 2018), which, among others, has an anti-naval surface unit strike role – a similar mission to that planned in the 1970's for the T-4/X-45 aircraft missile combination. The twenty first century Kinzhal, with its speed of up to Mach 10, must be regarded as invulnerable to modern air defence systems just as the planned Mach 7 capable X-45 would have been invulnerable to then modern air defence systems of the late twentieth century. It is a point of fact that a missile in the class of the X-45 would have been all but invulnerable to modern air defences if it were in service in the second decade of the twenty first century. In this regard, the Kinzhal air system, as had been planned more than four decades before with the T-4/X-45, bestows upon the Russian Federation a hypersonic conventional/nuclear strike capability against a diversity of target sets that is invulnerable to all known or projected air and missile defence systems.

The Kinzhal Air System consists of a MiG-31K launch platform and a Mach 10 capable aero-ballistic missile (Kh-47M2) that is similar in appearance to the aero-ballistic missile of the Iskander-M short-range ground-launched complex. The Kinzhal Air System entered limited service (10 launch platforms) with the Russian Federation Aerospace Forces in late 2017. The missile could be integrated with the Tu-22M3M. However, the Tu-22M3M is slated to be armed with the Mach 4-5 capable Kh-32, thought to be a deep modernisation of the Kh-22M2. MODRF

APPENDICES

Appendix I

Sukhoi T-4 specification – data furnished by JSC Sukhoi and the MODRF

Length of aircraft with pvd: 44.5 m
Height: 11.195 m
Wingspan: 22 m
Wing area: 295.7 m^2
Engines: Four x RD-36-41 afterburning turbojets each rated at ~16000 kg maximum thrust
Maximum speed at altitude: 3200 km/h
Practical ceiling: 24000 m
Practical range: 7500 km
Run-up/landing run: 1000/950 m
Weapons: Raduga X-45 (Kh-45) autonomously guided aero-ballistic hypersonic cruise missile (Sukhoi documentation states three such missiles could be carried while MODRF documentation states two)

Appendix II

T-4 Aircraft built or partially built

Aircraft	Function	First Flight	Status
Article 100S	Ground test	N/A	Scrapped
Article 101	Flight test	22 August 1972	Monino museum exhibit
Article 102	Flight test	N/A	Scrapped
Article 103	Flight test	N/A	Scrapped
Article 104	Flight test	N/A	Scrapped
Article 105	Flight test	N/A	Scrapped
Article 106	Flight test	N/A	Scrapped

NA = Not applicable
Note 1: Article 101 was the only flight test T-4 that was completed. Article 102 was almost complete when the program was terminated and the other airframes were in various degrees of construction
Note 2: Other aircraft were involved in the T-4 development program, including the Su-9FL 001 and other Su-9's employed as chase aircraft during flight testing

Appendix III

Flight Testing			
Phase	Number of flights	First Flight	Last flight
One	Nine	22.08.1972	08.08.1973
Two	One	19.01.1974	19.01.1974

Appendix IV

X-45 Design Team

A.Ya. Berezyak, G.K. Samohvalov, V.A. Larionov, A.I. Myakotin, B.I.O. Makov, R.Sh. Khaikin, O.V. Melnikov, A.N. Novikov, Y.I. Bazhanov, V.I. Belov, N.P. Mogutov

Appendix V

LABOROTORIES

Many of the various systems and design technologies were extensively tested in ground based and airborne laboratories prior to incorporation and testing on the actual T-4 flight article. These test programs were conducted under joint endeavours between Sukhoi Design Bureau and the various contractors for the respective systems and test organisations. The wing planform was extensively tested under a joint program between Sukhoi Design Bureau and the FRI using a modified Su-9 supersonic fighter fitted with a reduced scale version of the proposed T-4 wing planforms. This modified Su-9 flying laboratory was designated the 100L and was jointly operated by Sukhoi and the FRI (Sukhoi). Another Sukhoi fighter development, this time an Su-7, often referred to as the 100LDU, has been associated with the T-4 development program. While this is true, after a fashion, the 100LDU was in fact a general forward canard research aircraft

Appendix VI

Speed Regimes in Mach	
Subsonic	<Mach 1
Transonic	Mach 1
Supersonic	>Mach 1
Hypersonic	>Mach 5

Appendix VII

Mach (speed in relation to the speed of sound)

Mach = object speed over the speed of sound = Mach number

Mach 1 at sea level = 1125 km/h (761.1 mph/661 knots/340 m/s) with an air temperature of 288 K (15° C)

Mach 1 at 18288 m altitude = 1062 km/h (659.9 mph/573 knots/249 m/s) with an air temperature of 216 K (-56.6° C)

GLOSSARY

ASCM	Anti-Ship Cruise Missile
ASW	Anti-Submarine Warfare
C	Centigrade
FBW	Fly-By-Wire
FCS	Flight Control System
Ft.	feet (unit of measurement)
in.	inch (unit of measurement)
K	Kelvin
Kg	Kilogram
Km	Kilometer
Km/h	Kilometers per hour
knots	Nautical miles per hour
m	Metre (unit of measurement)
m^2	Metres squared
Mach	1 Mach – the speed of sound (this varies with altitude)
MAI	Minister of Aviation Industry
MiG	Mikoyan
MODRF	Ministry of Defence of the Russian Federation
Mph	Miles per hour
NASA	National Aeronautics and Space Administration
NATO	North Atlantic Treaty Organisation
NMUSAF	National Museum of the United States Air Force
OJSC	Open Joint Stock Company
RORSAT	Radar Ocean Reconnaissance Satellite
SR	Strategic Reconnaissance
SSB	Conventional powered ballistic missile submarine
SSBN	Nuclear powered ballistic missile submarine
SSN	Nuclear Powered Attack Submarine
SST	Supersonic Transport
Su	Sukhoi
TsAGI	Central Aerodynamic Institute
Tu	Tupolev
USAF	United States Air Force
USN HC	United States Naval Historical Center
USSR	Union of Soviet Socialist Republics
V/STOL	Vertical/Short Take-Off and Landing
XB	Experimental Bomber
~	Approximately equal to (can also be used to mean asymptotically equal)
<	Less than
>	Greater than
°	Degree(s)

ABOUT THE AUTHOR

Hugh Harkins, FRAS is a historian and author with an extensive background in astro/geophysics and studies/research in the wider scientific, aeronautic, astronautic and nautical technical and historical fields. Hugh has published in excess of sixty books; non-fiction and fiction, writing under his given name as well as utilising several pseudonyms. He has also written for several international magazines, whilst his work has been used as reference for many other projects ranging from the aviation industry, international news corporations and film media to encyclopaedias, museum exhibits and the computer gaming industry. Hugh is a member of the Institute of Physics and is an elected Fellow of the Royal Astronomical Society. He currently resides in his native Scotland. Other titles by the author include:

Iskander - Mobile Tactical Aero-Ballistic/Cruise Missile Complex
Orbital/Fractional Orbit Bombardment System - The Soviet Globalnaya Raketa
Counter-Space Defence Co-Orbital Satellite Fighter
Sukhoi T-50/PAK FA - Russia's 5th Generation 'Stealth' Fighter
Sukhoi Su-35S 'Flanker' E - Russia's 4++ Generation Super-Manoeuvrability Fighter
Sukhoi Su-34 'Fullback'
Sukhoi Su-30MKK/MK2/M2 - Russo Kitashiy Striker from Amur
MiG-35/D 'Fulcrum' F – Towards the Fifth Generation
Air War over Syria, Tu-160, Tu-95MS & Tu-22M3 - Cruise Missile and Bombing Strikes on Syria, November 2015-February 2016
Sukhoi Su-27SM(3)/SKM
Russian/Soviet Aircraft Carrier & Carrier Aviation Design & Evolution Volume 1 - Seaplane Carriers, Project 71/72, Graf Zeppelin, Project 1123 ASW Cruiser & Project 1143-1143.4 Heavy Aircraft Carrying Cruiser
Light Battle Cruisers and the Second Battle of Heligoland Bight
British Battlecruisers of World War 1 - Operational Log, July 1914-June 1915
Eurofighter Typhoon - Storm over Europe
Tornado F.2/F.3 Air Defence Variant
Air to Air Missile Directory
North American F-108 Rapier - Mach 3 Interceptor
Convair YB-60 - Fort Worth Overcast
Boeing X-36 Tailless Agility Flight Research Aircraft
X-32 - The Boeing Joint Strike Fighter
X-35 - Progenitor to the F-35 Lightning II
X-45 Uninhabited Combat Air Vehicle
Into The Cauldron - The Lancaster MK.I Daylight Raid on Augsburg
Hurricane IIB Combat Log - 151 Wing RAF, North Russia 1941
RAF Meteor Jet Fighters in World War II, an Operational Log
Typhoon IA/B Combat Log - Operation Jubilee, August 1942
Defiant MK.I Combat Log - Fighter Command, May-September 1940
Blenheim MK.IF Combat Log - Fighter Command Day Fighter Sweeps/Night Interceptions, September 1939 - June 1940
Tomahawk I/II Combat Log - European Theatre, 1941-42
Fortress MK.I Combat Log - Bomber Command High Altitude Bombing Operations, July-September 1941
XF-92 - Convairs Arrow

www.ingramcontent.com/pod-product-compliance
Lightning Source LLC
Chambersburg PA
CBHW052042190326
41519CB00003BA/255